RULERS OF THE OCEANS

John Allan

Picture Credits
(abbreviations: t = top; b = bottom; m = middle; l = left; r = right; bg = background)

Shutterstock: Alessandro De Maddalena 2tl, 26-27bg, 29bl; Alexyz3d 31mr; Andrei Armiagov 31tl; Catchlight Lens 31bl; DiveSpin.Com 18-19bg; Fon Duangkamon 22-23bg, 28tr; Foto 4440 30ml; Jayaprasanna T.L 7m; JENG BO YUAN 8-9bg; JMx Images 7t; kaschibo 14-15bg; layton harrison 20-21bg, 29tl; R raymoonds 16-17bg; RobJ808 10-11bg, 28b; simongee 3b, 6bl, 12-13bg, 28tl; cSubphoto.com 6m; Tory Kallman 24-25bg, 29mr; YUSRAN ABDUL RAHMAN 30br.

Every effort has been made to trace the copyright holders and we apologise in advance for any unintentional omissions. We would be pleased to insert the appropriate credit in any subsequent edition of this publication.

Copyright © 2025 Hungry Tomato Ltd

First published in 2025 by Hungry Tomato Ltd
F15, Old Bakery Studios, Blewetts Wharf, Malpas Road, Truro, Cornwall, TR1 1QH, UK.

No part of this publication may be reproduced, stored in a retrieval system, or transmitted in any form or by any means, electronic, mechanical, photocopying, recording, or otherwise, without prior written permission of the copyright owner.
A CIP catalogue record for this book is available from the British Library.

ISBN 9781835690765

Printed in China

Discover more at
www.hungrytomato.com

RULERS OF THE OCEANS

MEET THE WORLD'S MOST DANGEROUS SEA CREATURES!

CONTENTS

Rulers Of The Oceans	6	Nile Crocodile	20
Stonefish	8	Box Jellyfish	22
Stingray	10	Great White Shark	24
Cone Snail	12	Orca	26
Blue-Ringed Octopus	14	When Predators Become Prey	28
Piranha	16	Fearsome Facts	30
Beaked Sea Snake	18	Glossary & Index	32

Words in **BOLD** can be found in the glossary.

RULERS OF THE OCEANS

Which animal wins the title of deadliest ruler of the ocean? This big question isn't as easy to answer as you might think! There's lots to consider...

DEADLY IN THE DEEP BLUE

This book includes some of the most dangerous sea creatures on Earth, from the dark depths, right up to the ocean's surface!

DEADLY CREATURES

The deadliest animals come in all shapes and sizes. Explore the top ten deadly defenders of the deep, from big, sharp-toothed crocodiles to small (but vicious!) piranhas...

WHAT'S FOR DINNER?

Carnivores are animals that only eat meat, herbivores are animals that only eat plants, and omnivores are animals that eat both! Most of the extreme **predators** we explore in this book are carnivores, which makes them the most dangerous of all!

DEADLY COUNTDOWN

Each animal is ranked in order from the 10th most deadly, down to the ultimate underwater predator... it's not always the biggest that win!

WARNING

THINGS GET GRIM FROM HERE ON IN... TURN THE PAGES TO FIND OUT MORE!

STONEFISH

Meet the fish that looks just like a rock! Stonefish hide in wait for unsuspecting reef fish, which they like to eat. Not only are they very unusual to look at, stonefish are also the most **venomous** fish in the sea!

STONE OR FISH?
This predator is great at blending in! The stonefish has lumpy skin that acts as a natural **camouflage** - it changes its appearance to match the habitat it's in. The **prey** of a stonefish never see them coming!

SHARP ATTACK

The stonefish will lie close to the **seabed** to hide from unsuspecting prey. There, it will wait for its prey to swim by, before using its sharp spines to attack! The stonefish's spines easily stab through flesh and contain deadly **venom** that can kill quickly.

FACT FILE

WEIGHT
Up to 2 kg
(5 lbs)

DIET
Carnivore

LOCATION
Indian and
Pacific Oceans

LETHAL POWERS
Sharp spines, deadly venom, and great camouflage skills

DEADLY COUNTDOWN

NO.10

STINGRAY

Stingray are closely related to sharks, but you don't have to worry about their bite - it's the sting in the tail that's the problem! These flat fish are seabed dwellers, and often bury themselves under the sand in shallow waters – swimmers beware!

A STINGY SITUATION

A key feature of the stingray is its powerful spiked tail, which can jab an attacker with painful venom. Some stingrays can even create an electric charge, shocking prey so they can quickly eat them with their crushing jaws.

BONELESS BODIES

Stingrays have no bones! Their skeletons are made up of bendy **cartilage** which allow them to move their bodies quickly across the waves when chasing prey.

FACT FILE

WEIGHT
Up to 365 kg (800 lbs)

DIET
Carnivore

LOCATION
Pacific and Atlantic Oceans

LETHAL POWERS
Spiked tail, deadly venom, and flexible body

DEADLY COUNTDOWN

NO.9

CONE SNAIL

If you saw one of these on the beach you might be tempted to pick it up – bad move! Cone snails can cause a painful **sting**, with some **species** releasing deadly **toxins** that paralyse their prey. Found in tropical **coral reefs**, a cone snail's pretty exterior is the perfect camouflage.

BULLS-EYE!
Cone snails have four tubes that poke out the front of their shell. These include two eyestalks, a sensor that helps them detect prey, and the **proboscis**, a lethal dart that can inject paralysing toxins!

TINY BUT DEADLY
Cone snails are carnivorous **mollusks**. They usually hunt at night, using their proboscis to trap their food. They eat their prey using a row of tiny teeth called the **radula**.

FACT FILE

WEIGHT
Up to 2 kg
(4.5 lbs)

DIET
Carnivore

LOCATION
Indian and
Pacific Oceans

LETHAL POWERS
Deadly venom, paralysing toxins, and spiked darts

DEADLY COUNTDOWN
NO.8

BLUE-RINGED OCTOPUS

The blue-ringed octopus is not only one of the most beautiful of sea creatures – it's also one of the most deadly! This small, shy animal can be found in tide pools and coral reefs. Swimmers have learned not to go looking for it because it has a very nasty bite!

POWERFUL BITE
The blue-ringed octopus is the only octopus that has a venomous bite. Before killing it, the venom makes prey unable to move their body. What a nasty attack!

EASY PREY

This octopus mostly feeds on crabs and wounded fish that can't get away. It has a sharp beak that makes it easy to slice through flesh.

FACT FILE

WEIGHT
Up to 80 g (2.9 oz)

DIET
Carnivore

LOCATION
Indian and Pacific Oceans

LETHAL POWERS
Quick, venomous bite, and a sharp beak

DEADLY COUNTDOWN

NO.7

15

PIRANHA

This fish has earned its nickname as the 'Wolf of the Water'! Piranhas have sharp teeth, hunt in large groups, and have a big appetite for meat. One piranha can give you a nasty bite, but a group of hungry piranhas can eat pretty much any animal!

ALIVE PREY
Piranhas are lethal predators. They don't just kill their prey; they start eating it alive!

SUPER SENSES

Piranhas also have a super sense of smell and can detect blood in the water, even if it's far away. They may not often attack humans, but they can be nasty, so watch where you swim!

FACT FILE

WEIGHT
Up to 3.6 kg (8 lbs)

DIET
Omnivore

LOCATION
South America

LETHAL POWERS
Aggressive, super sense of smell, and very sharp teeth

DEADLY COUNTDOWN

NO.6

BEAKED SEA SNAKE

The beaked sea snake is a bad-tempered killer. This one species is responsible for over half of all sea snake attacks on humans! Its venom is deadlier than most other creatures on land. They feed mostly on fish, including eels.

LOOK OUT!
The vivid black markings on the beaked sea snake make it easy to identify. They have a special flat tail for swimming and thin skin over their nostrils which close when in water.

FACT FILE

WEIGHT
Up to 1.3 kg
(3 lbs)

DIET
Carnivore

LOCATION
Philippines and
North Australia

LETHAL POWERS
Deadly venom, **agile**, and
very aggressive

POISONOUS HUNTERS

These predators hunt by injecting venom into their prey's muscles, and stopping them from breathing. The beaked sea snake can swallow prey twice the size of its neck!

DEADLY COUNTDOWN

NO.5

NILE CROCODILE

The Nile crocodile is a cold-blooded killer, but can be rather lazy when it comes to hunting. This large reptile prefers to lie in water with only its eyes and nostrils showing, waiting for prey to come close enough to attack.

BIG HUNTERS

Like other reptiles, the Nile crocodile can bite, but not chew. This means all of its prey is swallowed whole, no matter how big. Nile crocodiles can even take on giraffes! Long jaws full of sharp teeth are this predator's main weapon.

FACT FILE

WEIGHT
Up to 750 kg
(1,650 lbs)

DIET
Carnivore

LOCATION
Africa

LETHAL POWERS
Long jaws, sharp teeth, and very aggressive

DEADLY COUNTDOWN

NO.4

SCALY SKIN
This fierce predator is covered in bony plates that keep it safe when taking on big prey that might fight back! The Nile crocodile moves fast in the water, but can travel twice as quickly on land!

BOX JELLYFISH

Box jellyfish, also known as sea wasps, have developed one of the most deadly venoms in the world. They have lots of stinging tentacles holding venom that kills almost instantly. To make it worse, this jellyfish is transparent, meaning it's almost impossible to spot under water!

FREAKY FEATURES

This jellyfish has a clear box-shaped body, and can grow tentacles up to 3 metres (10 ft) long, perfect for reaching out to sting its victims! It is known for having excellent eyesight, and can spot danger from far away. This jellyfish can attack a predator before it sees them!

DEADLIEST JELLYFISH

Each tentacle has sharp, sensitive points that inject deadly venom at the smallest of touches. When this jellyfish stings its victims, it can stop their heart from beating in just a few seconds! To protect people, many beaches close when there are box jellyfish around.

FACT FILE

WEIGHT
Up to 2 kg (4 lbs)

DIET
Carnivore

LOCATION
Australia

LETHAL POWERS
Hard to spot, deadly venom, and stinging tentacles

DEADLY COUNTDOWN
NO.3

GREAT WHITE SHARK

The great white shark is one of the most well-known predators in the world, not just in the ocean. This deadly killer mostly eats seals, dolphins, and other sharks, but will attack anything it thinks it can eat, including humans!

SUPER SMELL
The great white shark has a fantastic sense of smell, helping it find wounded prey from far away.

TERRIFYING ATTACKER

This fierce predator attacks prey with a twisting lunge, tearing a chunk of flesh off. Then, it waits for the victim to die before coming back to eat! Its rows of razor-sharp teeth slice through flesh easily, making it impossible for prey to escape once attacked.

FACT FILE

WEIGHT
Up to 1,815 kg (4,000 lbs)

DIET
Carnivore

LOCATION
Every ocean

LETHAL POWERS
Rows of sharp teeth, very aggressive, and fantastic sense of smell

DEADLY COUNTDOWN
NO.2

ORCA

The orca lives up to its nickname, 'killer whale'. They are deadly predators with no natural enemies. Orcas live and hunt in family groups, and are known for their brutal hunting techniques. These skilled predators take the number one spot in the deadly countdown!

HUNGRY PREDATORS

Orcas have up to 50 large, pointed teeth, but can't chew, so they have to swallow their prey whole – no matter the size!

CLEVER HUNTERS

Orcas are very intelligent creatures, working together to find the best way to attack prey. Some use their tails to smack prey, others surround prey from all directions, and some even flip seals off icebergs!

FACT FILE

WEIGHT
Up to 7,260 kg (16,000 lbs)

DIET
Carnivore

LOCATION
Every ocean

LETHAL POWERS
Highly intelligent, large teeth, and hunts in groups

DEADLY COUNTDOWN
NO.1

WHEN PREDATORS BECOME PREY

These deep diving predators are some of the deadliest in the animal kingdom. But what happens when these lethal hunters are hunted themselves?

STINGRAY
Sadly, stringrays are threatened wih **extinction**. This is due to overfishing, **climate change**, and a loss of their habitat.

CONE SNAIL
Cone snails are at risk because they are over-hunted for their highly-prized shells! They are also collected for scientific research because scientists think their venom could make a powerful pain-killing medicine.

28

BLUE-RINGED OCTOPUS

While extremely poisonous to humans, some animals such as eels, dolphins, and seals are not affected by blue-ringed octopus' venom. This means this octopus often becomes lunch for these animals when they get hungry!

NILE CROCODILE

Very little threatens the adult Nile crocodile, which means they are at the top of the food chain! However, eggs and even baby Nile crocodiles are sometimes stolen and eaten by lizards and monkeys.

FEARSOME FACTS

There are so many deadly and fearsome facts about each ruler of the ocean. Here are some more that show just how impressive these predators really are...

ORCA
Orcas will often swim onto beaches in order to hunt seals and sea lions.

BOX JELLYFISH
Box jellyfish may be deadly, but they do not have a brain!

STINGRAY
Stingray can defend and hunt for themselves as soon as they are born!

BLUE-RINGED OCTOPUS
Dead or alive, the venom of a blue-ringed octopus is poisonous to humans.

PIRANHA
They can live in protective packs of about 1,000 fish to defend against predators.

STONEFISH
This tough little predator can survive a whole 24 hours out of water!

BEAKED SEA SNAKE
This snake's long beak is used to open up cracks in coral reefs and capture hiding prey.

GREAT WHITE SHARK
These sharks can propel themselves through the water at 35 miles (60 km) per hour!

NILE CROCODILE
Unusual in reptiles, both mother and father crocodiles will guard their nest until eggs hatch.

CONE SNAIL
Despite their small size, cone snails are one of the fastest known hunters in the animal kingdom.

GLOSSARY

Agile - able to move quickly and easily.

Camouflage - when an animal blends into its surroundings so it's hard to see.

Cartilage - a flexible material that makes up part of many animals' bodies, including humans. We have cartilage in places like our nose and ears.

Climate change – a change in the weather conditions over a long time.

Coral Reefs - a rocky area in warm, shallow seas that is made of the chalky remains of tiny animals called coral polyps.

Extinction – when something like a plant or animal species no longer exists.

Mollusks - animals with a soft body and no backbone. They often have hard shells.

Prey - an animal that is hunted and killed by other animals for food.

Proboscis - a long moveable nose, like an elephant's trunk.

Radula - a band of tiny teeth found in the mouth of mollusks (see left).

Seabed - the bottom of the ocean.

Species - a group of living things that have the same features as each other and share a common name.

Sting - a sharp, piercing part of an animal often ejecting a venomous substance.

Toxins - substances that can cause harm or injury to living organisms.

Venom - a poisonous substance of an animal, usually passed on by a bite or sting.

Venomous - a creature that can produce venom (see above).

INDEX

B
Blue-ringed octopus 14-15, 29, 30
Beaked sea snake 18-19, 31
Box jellyfish 22-23, 30

C
Carnivore 7, 9, 11, 13, 15, 19, 21, 23, 25, 27
Cone snail 12-13, 28, 31

G
Great white shark 24-25, 31

H
herbivore (plant-eater) 7

N
Nile crocodile 20-21, 29, 31

O
omnivore 7, 17
Orca (killer whale) 26-27, 30

P
Piranha 6, 16-17, 31

S
Stingray 10-11, 28, 30
Stonefish 8-9, 31